AF305186

NORTH WARNING SYSTEM

Donovan Wylie

NORTH WARNING SYSTEM

Steidl

Cape Kakiviak. Northern Labrador

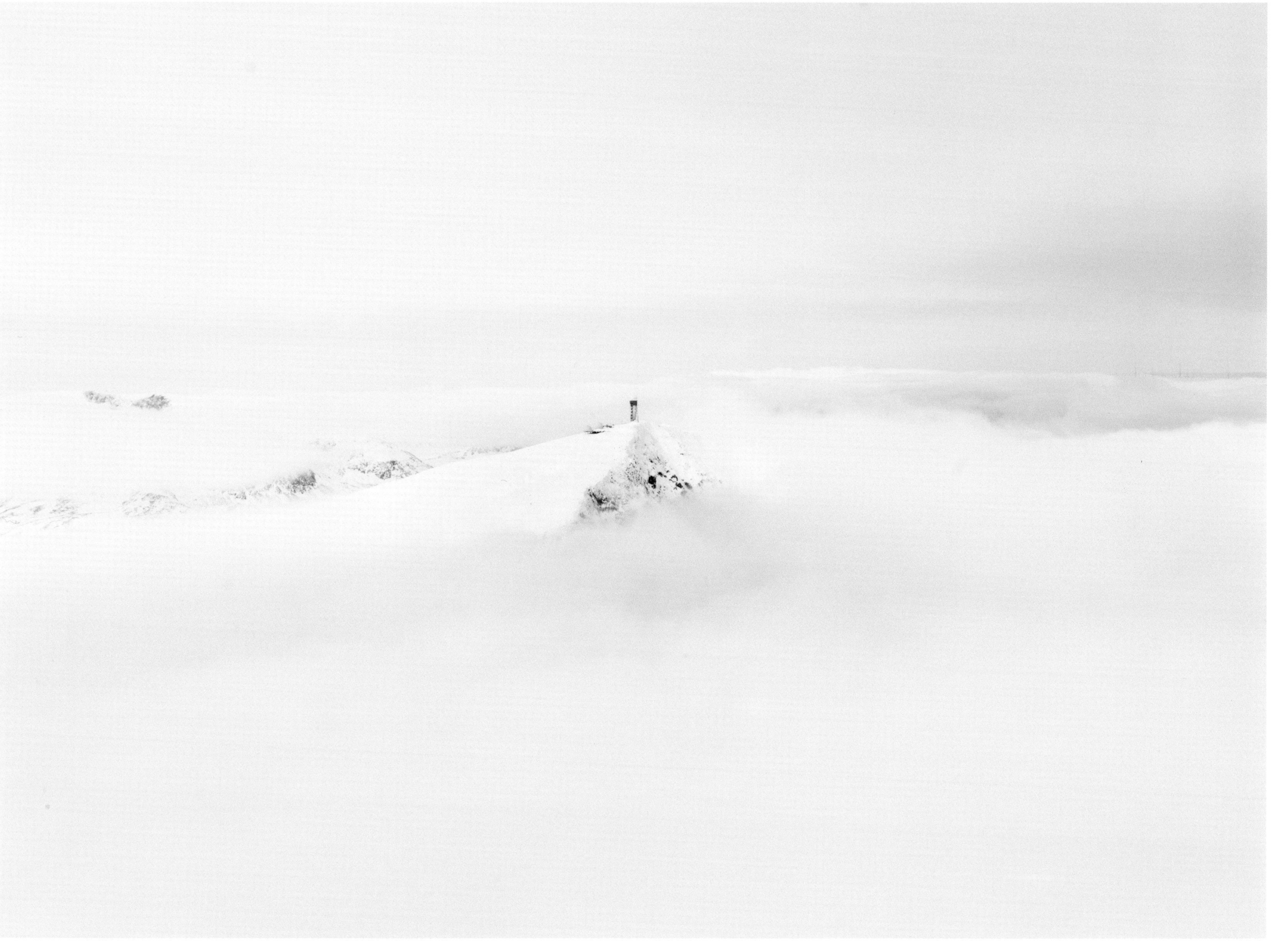

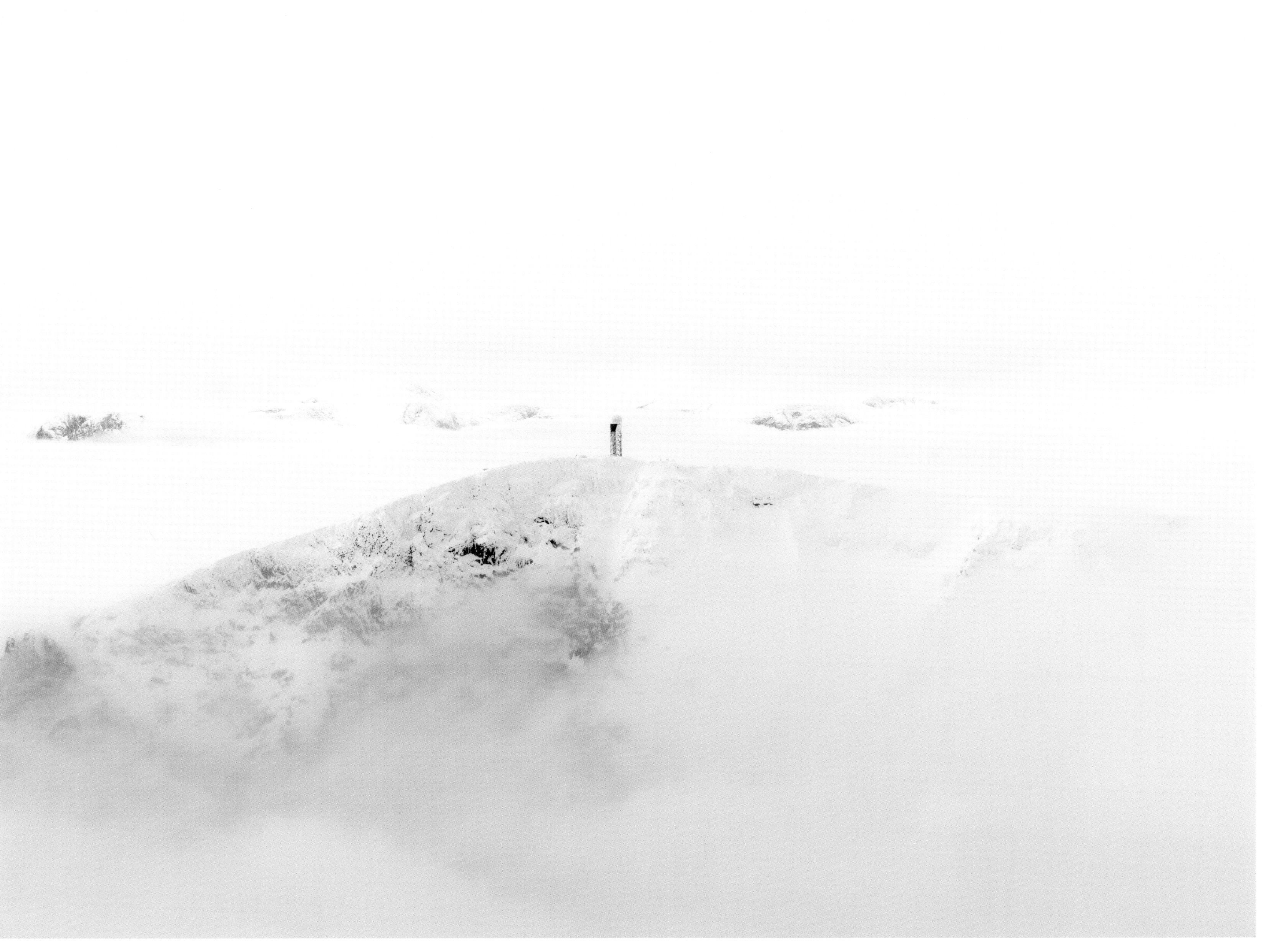

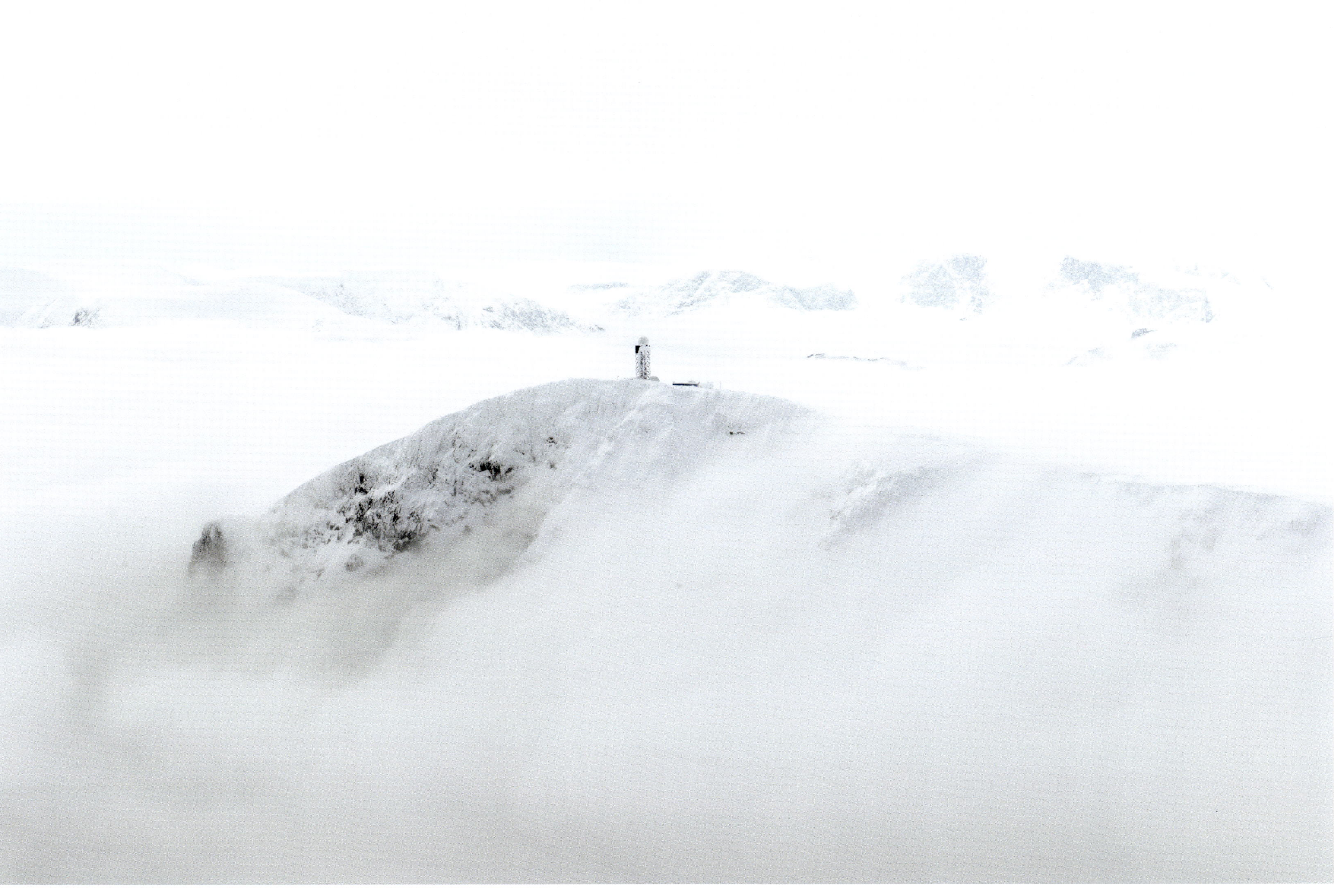

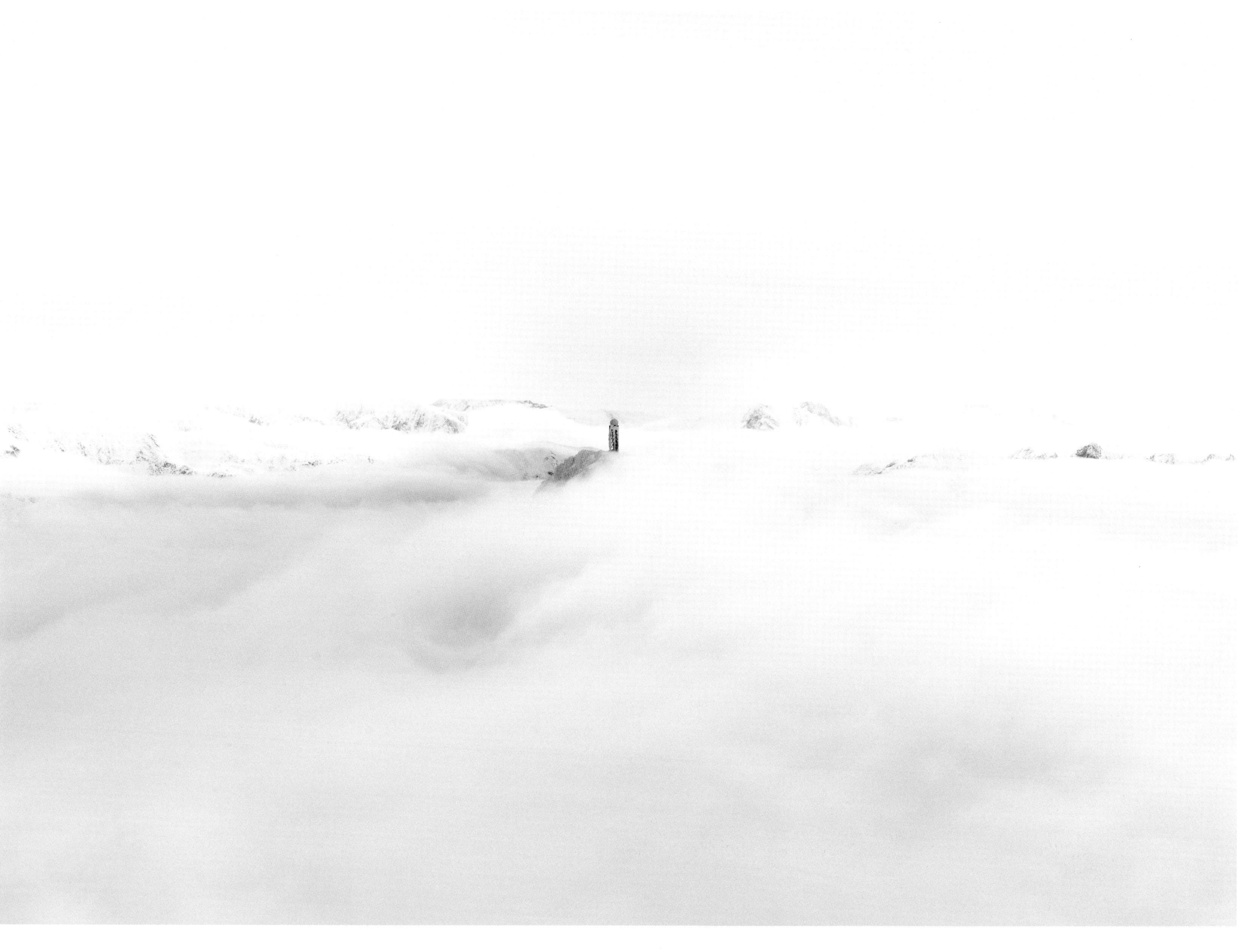

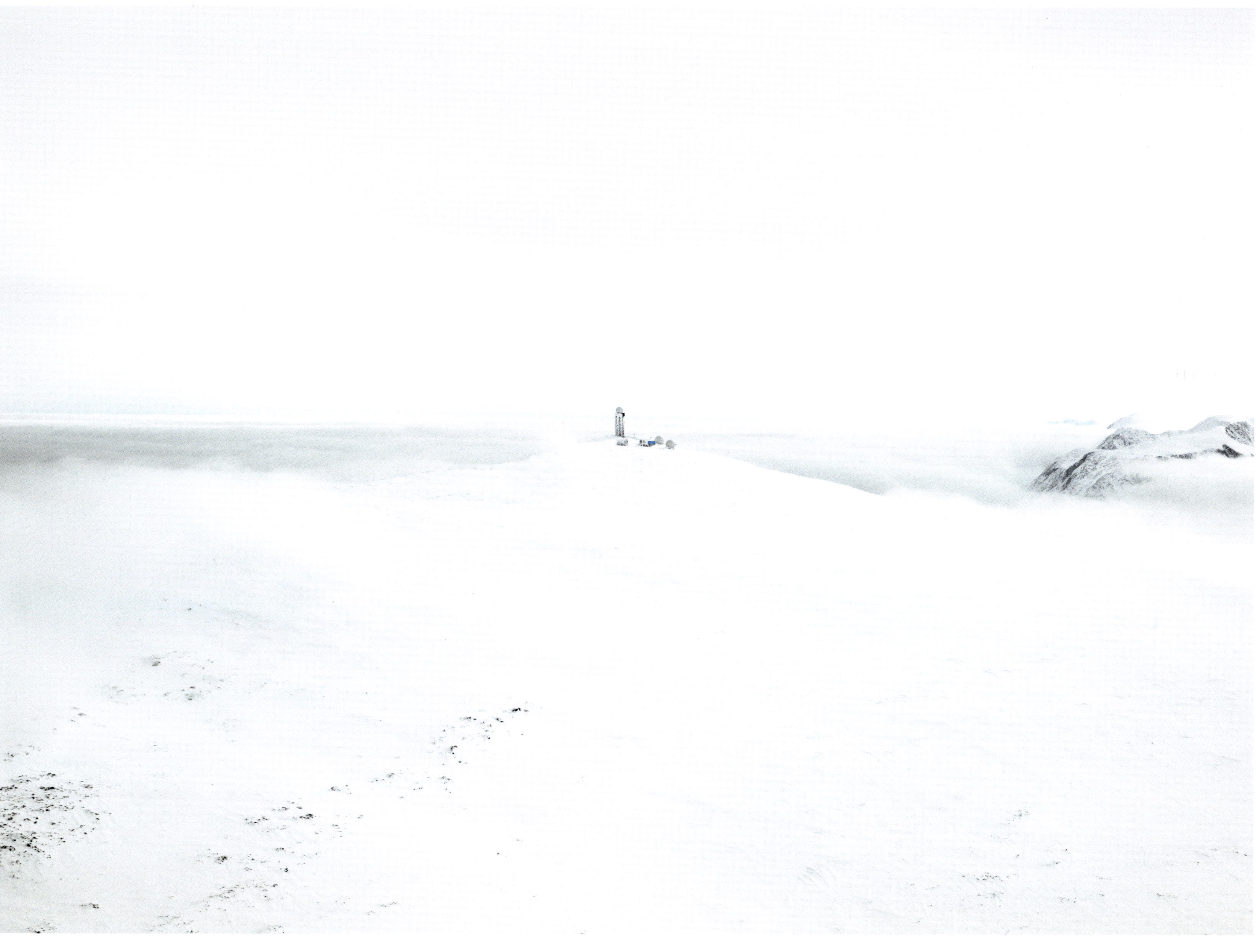

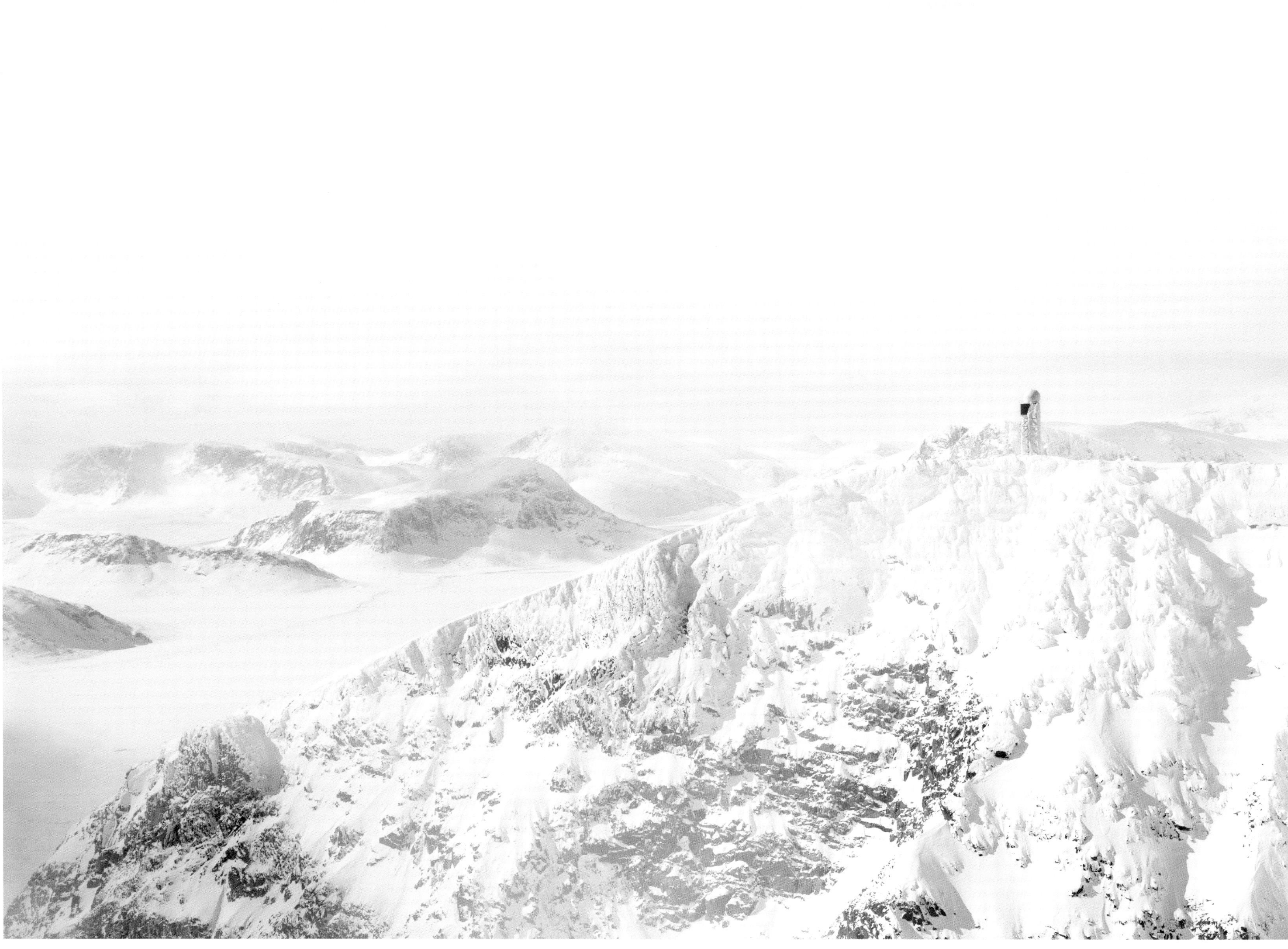

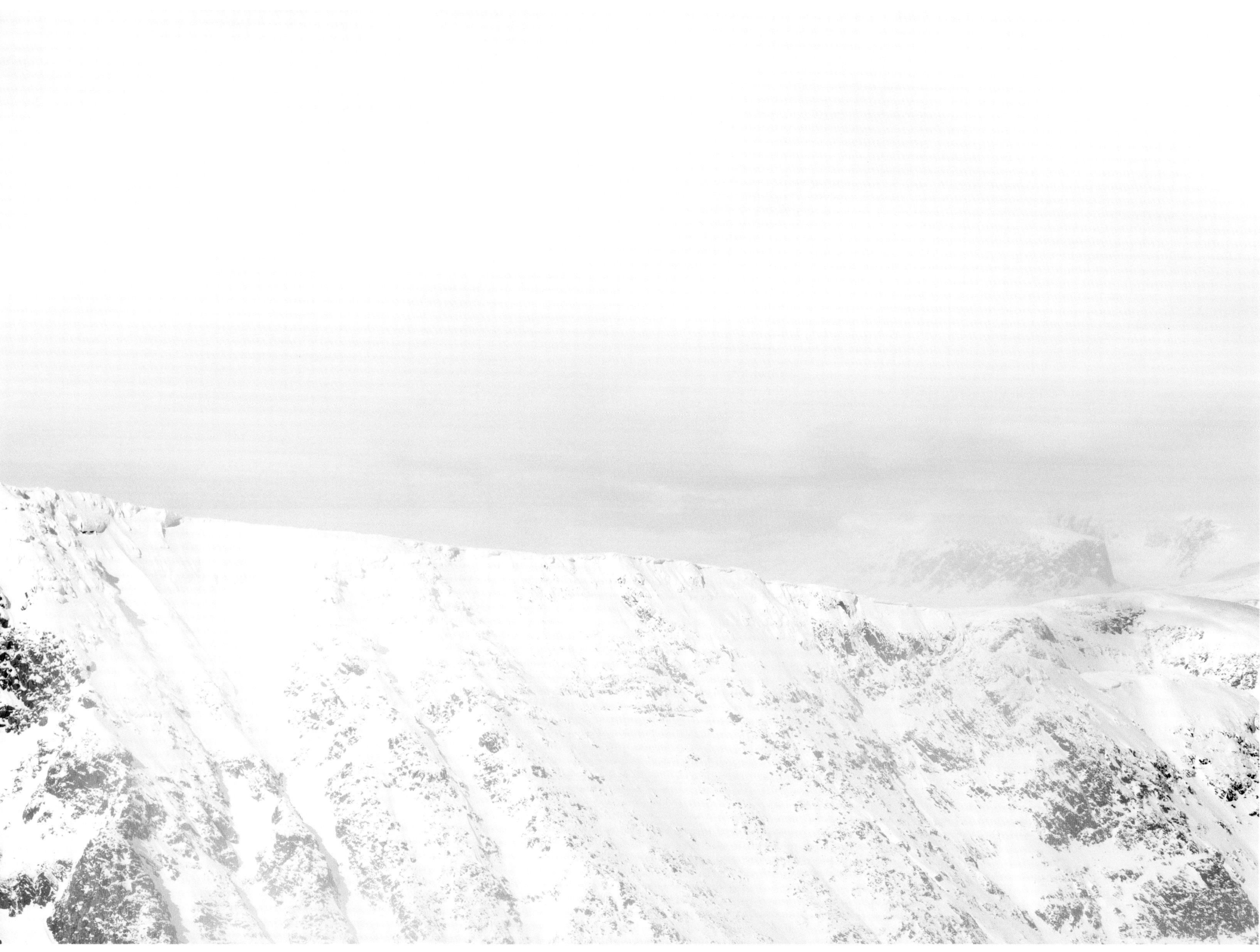

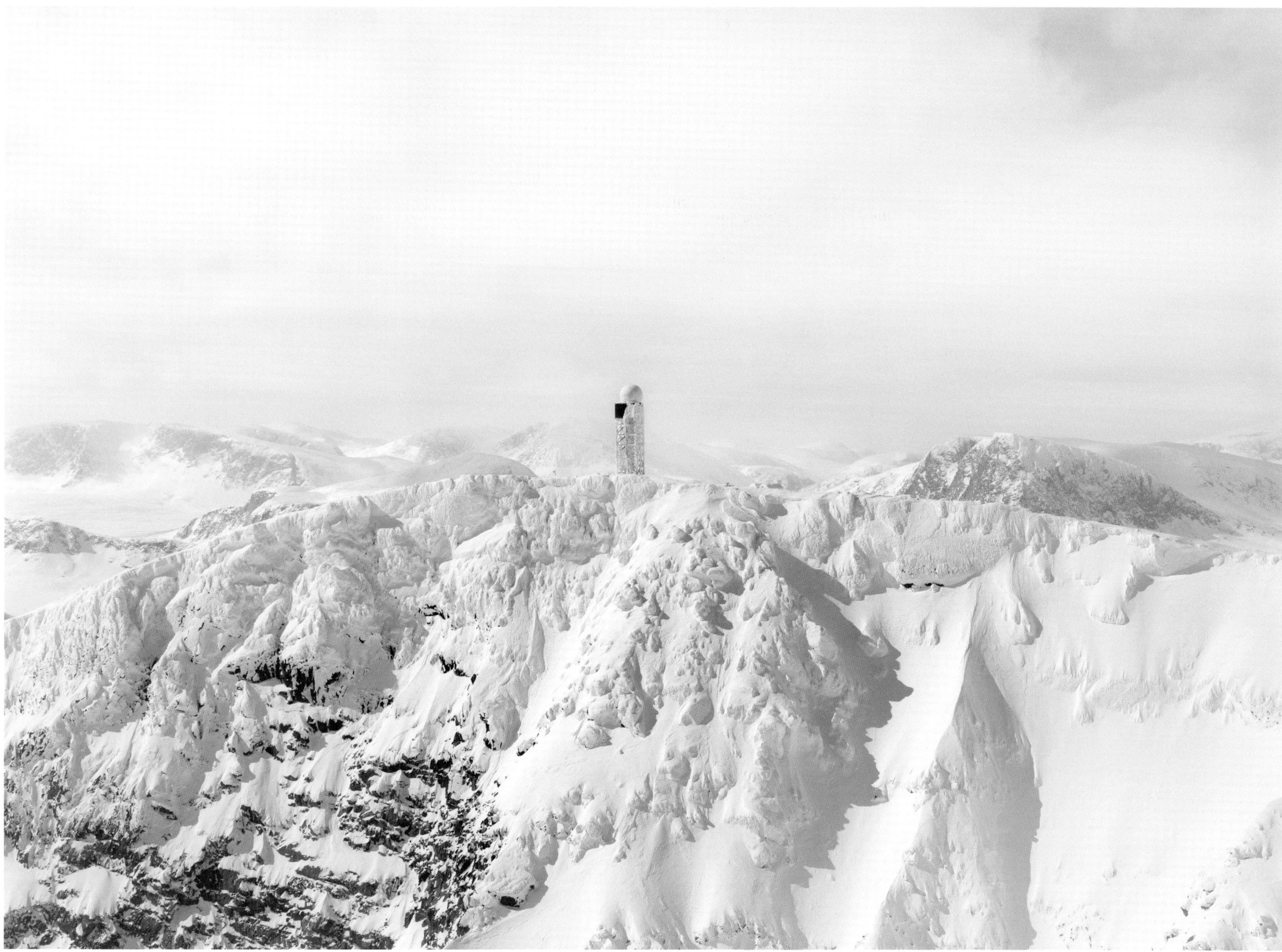

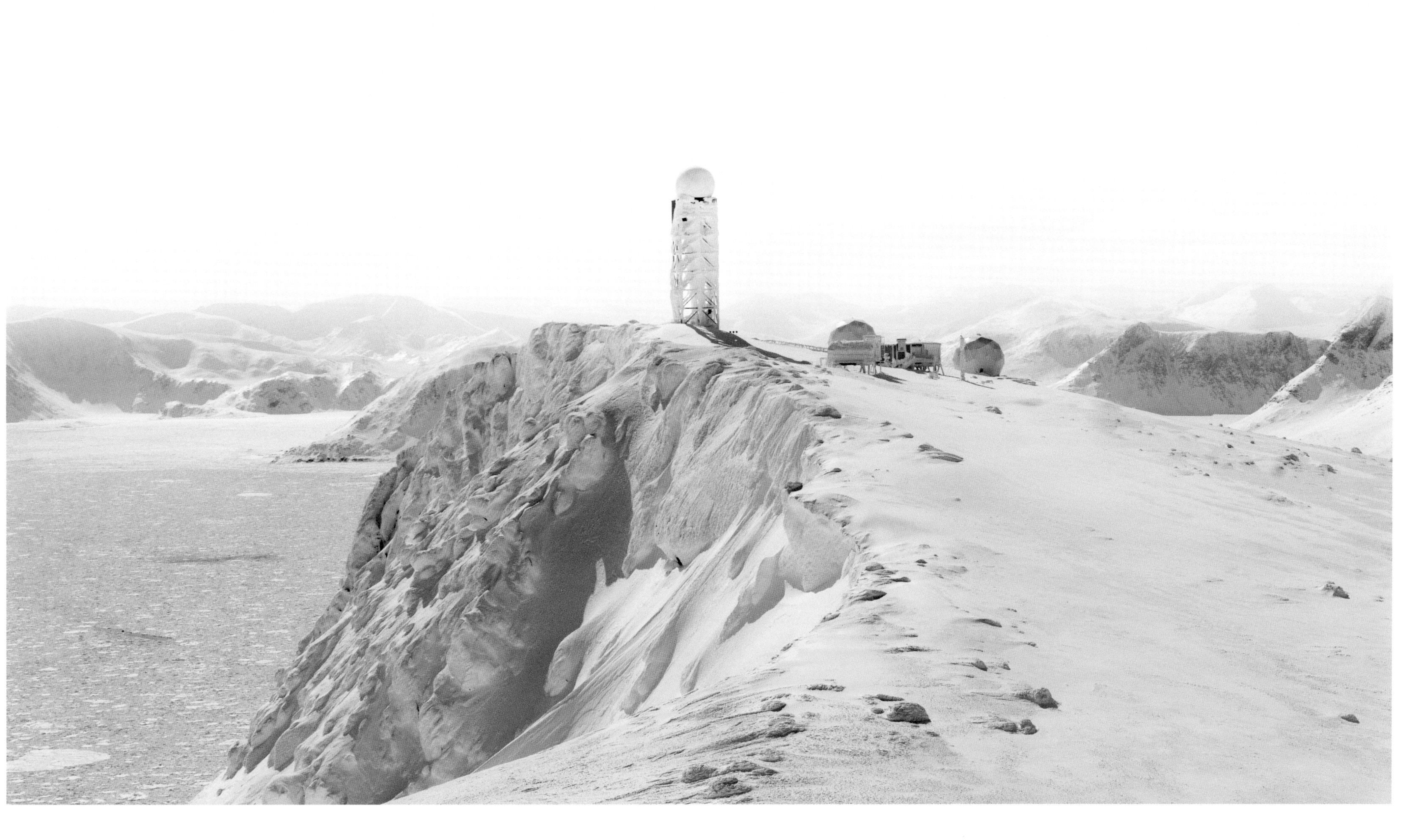

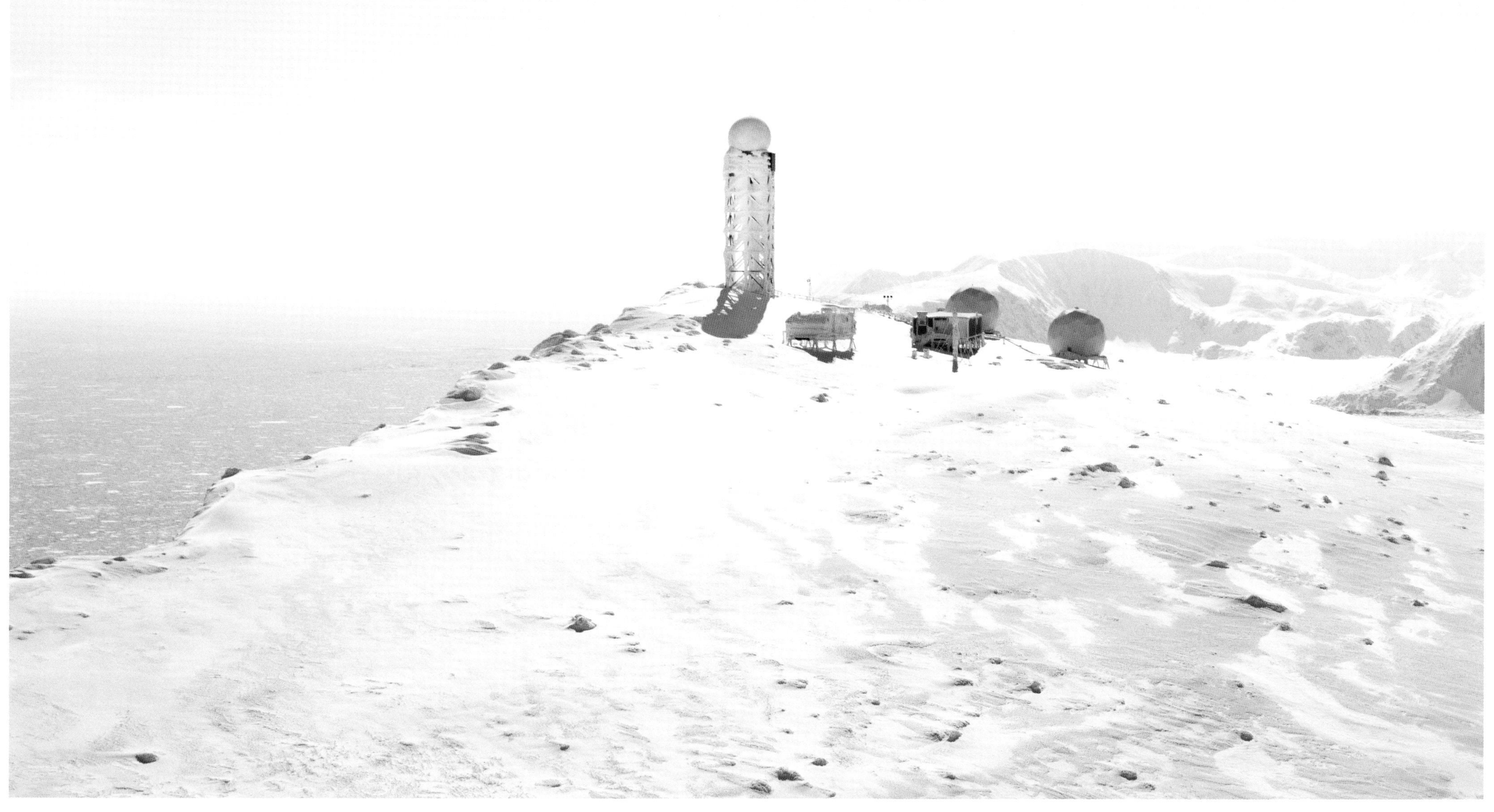

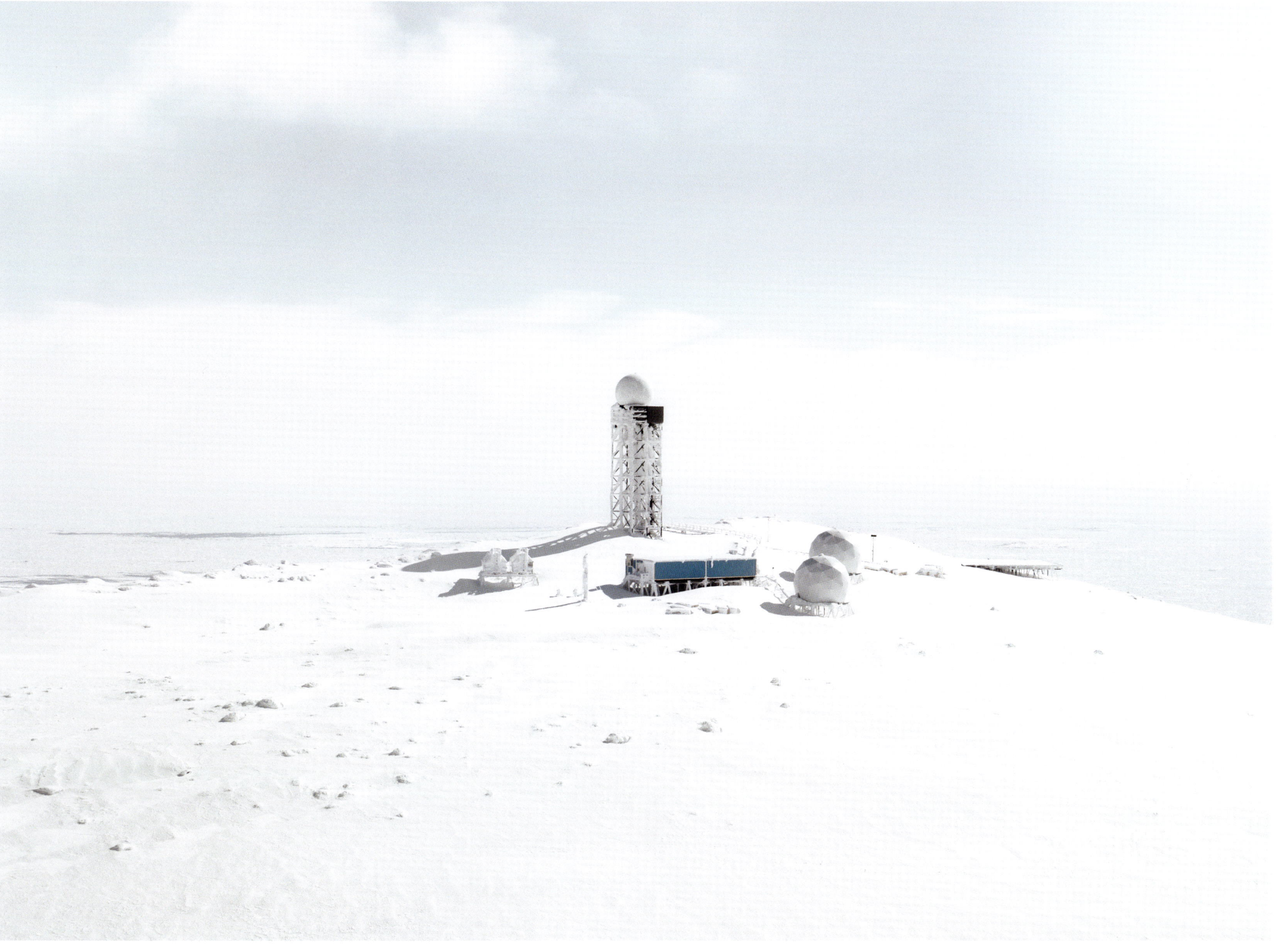

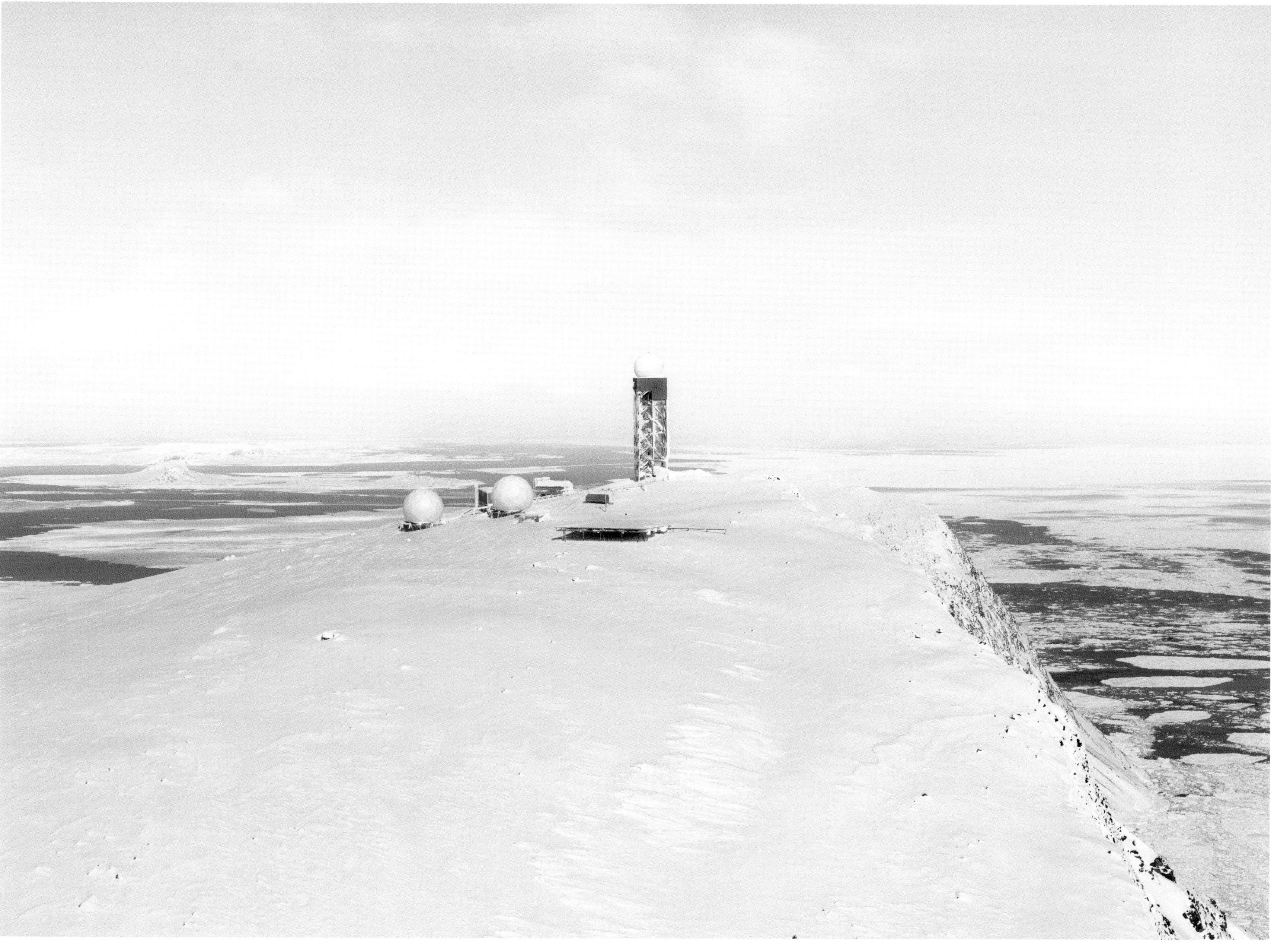

North Warning System

The development of long-range bombers and missiles after the Second World War made Canada's arctic frontier vulnerable to attack from the air. This forced Canada and the United States jointly to construct a matrix of short- and long-range radar stations in the 1950s. Known as the Distant Early Warning Line, these stations provided electronic observation and surveillance capability across Canada's northern frontier throughout the Cold War. In the 1990s, these stations were upgraded to form the North Warning System (NWS). The North Warning System is increasingly active – as international maritime traffic is developing throughout the north, so is military presence.

The whiteness of the arctic frontier, in the context of this work, represents the idea of a blank canvas. It is a metaphor for the sweep of history; the whiteness is a *"what if…"* The station operates electronically and seeks to detect an invisible threat in an empty landscape. After a 600km journey across the northern coasts of Quebec and Labrador, the helicopter I was in circled the radar station – Cape Kakiviak or "Lab1". The radar surveyed my presence and the passage, its scripted radius. I operated my own camera as mechanically and systematically as humanly possible while photographing the radar, mirroring its own rotational sweep that penetrates the surrounding expanse of land and sea. Once, for a fraction of a second, I released my focus and just looked. This tiny, forceful human presence in the blinding white seemed as unconscious and irrepressible as any ant colony or bee hive.

Donovan Wylie

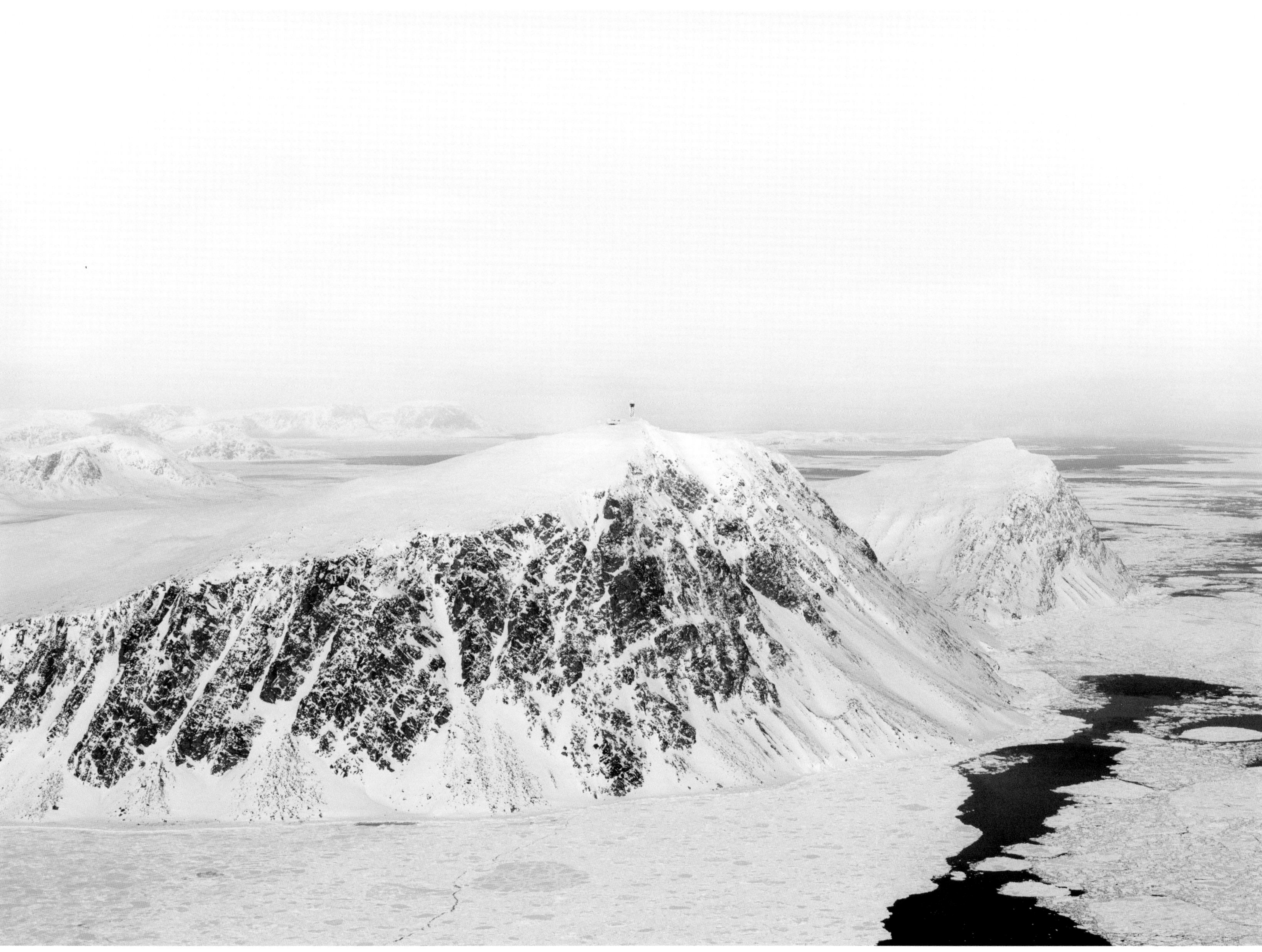

First edition published in 2014

Book design: Donovan Wylie, Gerhard Steidl, Bernard Fischer
Separations by Steidl's digital darkroom
Production and printing: Steidl, Göttingen
Steidl / Düstere Str. 4 / 37073 Göttingen, Germany
Phone +49 551 49 60 60 / Fax +49 551 49 60 649
mail@steidl.de / www.steidl.de
ISBN 978-3-86930-773-2 / Printed in Germany by Steidl